I0504220

POCKET
PHYSICS THEORY
THINGS YOU SHOULD KNOW
(QUESTIONS AND ANSWERS)

By Rumi Michael Leigh

Introduction

I would like to thank and congratulate you for downloading this book, *"Pocket Physics Theory, things you should know (questions and answers)"* series.

This book will give you a good general knowledge about the essentials of Physics Theory.

Thanks again for downloading this book, I hope you enjoy it!

General: Part 1: Questions

1. What is micro?
2. What is mega?
3. What is giga?
4. What is nano?
5. What is pico?
6. What is a scientific notation?
7. What is sin θ?
8. What is cos θ?
9. What is tan θ?

General: Part 1: Answers

1. 10-6.
2. 10 6.
3. 10 9.
4. 10-9.
5. 10-12.
6. This is a way of representing numbers using powers of ten.
7. Opposite/hypotenuse.
8. Adjacent/hypotenuse.
9. Opposite/adjacent.

Energy: Part 2: Questions

1. What is calorie?
2. What is work?
3. What is the unit of work?
4. What is energy?
5. What is kinetic energy?
6. What is potential energy?
7. What is power?

Energy: Part 2: Answers

1. This is the amount of heat required to raise the temperature of a gram of water by 1 degree.
2. Work is force applied at a certain distance.
3. Joules.
4. Energy is the ability to do work.
5. Kinetic energy is the energy of movement.
6. Potential energy is the energy stored that can be transformed to do work.
7. Power is the work done over a given time.

Dynamics: Part 3: Questions

1. What is speed?
2. What is instantaneous speed?
3. What is average velocity?
4. What is terminal velocity?
5. What is average acceleration?
6. Can an object in equilibrium be in motion?
7. What is free fall?
8. What is displacement?

Dynamics: Part 3: Answers

1. Speed is the distance travelled over time.
2. This is the speed in the moment.
3. This is the change in position of an object over a change in time.
4. This is the highest velocity reached by a free-falling object.
5. This is the change in velocity over a change in time.
6. Yes, but it's velocity is constant.
7. This is the absence of friction when an object is falling.
8. This is the difference between the final position of an object and its initial position.

Electricity: Part 4: Questions

1. What is a circuit?
2. Name 2 types of connections.
3. What is a series connection?
4. What are parallel connections?
5. Does the current in a series circuit differ?
6. What is an open circuit?
7. What is a closed circuit?
8. What happens when there are two electrical charges of the same sign?
9. What happens when there are two electrical charges of opposite signs?
10. What is an electric current?

Electricity: Part 4: Answers

1. A circuit is a path where electrons flow.
2. Series connections and parallel connections.
3. This is a system of connections in which current flows in one path.
4. This is a system of connections in which current flows in more than one path.
5. No, it is the same everywhere.
6. This is a circuit in which no current can flow.
7. This is a circuit in which there is a continuous flow of current.
8. They repel each other.
9. They attract each other.
10. An electric current is the flow of charge from one point to another.

Electricity: Part 5: Questions

1. What is the function of an electroscope?
2. What is the function of an electrometer?
3. What is the function of a capacitor?
4. What is a direct current DC?
5. What is an alternating current?
6. What is the relationship between the resistance in a circuit and current?
7. What is a photovoltaic cell?
8. Does a positively charged body have an excess of electrons?
9. Are all the electrons of a metal free?
10. What is used to measure the voltage in a circuit?

Electricity: Part 5: Answers

1. It is used for the detection of static electricity.
2. It is used to measure electric charges.
3. It is a device that stores potential energy.
4. This is a current that flows in one direction.
5. This is a current that periodically reverses their direction.
6. A current is high if the resistance is low.
7. A photovoltaic cell transforms light energy to electrical energy.
8. No.
9. No.
10. A voltmeter.

Electricity: Part 6: Questions

1. What is static electricity?
2. What is a superconductor?
3. What is induction?
4. What is a resistor?
5. What is a fuse?
6. What is a galvanometer?
7. What is the function of an electric motor?
8. Define electrostatics.
9. Give an example of a semiconductor.
10. What are transistors?
11. What is a diode?
12. What is an ammeter?

Electricity: Part 6: Answers

1. This is electricity created by friction.
2. This is a substance that conducts electricity without resistance.
3. This is the passing of electric or magnetic current between objects without any physical contact between the objects.
4. This is a device that resists electric current.
5. This is an electrical device that protects overheating by interrupting the flow of electric current.
6. This is an instrument used for measuring small electric current.
7. It converts electrical energy to mechanical energy.
8. This is the study of static electricity. Electricity is at rest.
9. Silicon.
10. Transistors are semiconductors that regulate current and are used to switch electronic signals.
11. This is a two terminal electronical device that allows the flow of current in one direction.

12. This is an instrument used to measure electric current.

Light and Optics: Part 7: Questions

1. What is the law of reflection?
2. What is a ray?
3. What is umbra?
4. What is penumbra?
5. What is the critical angle?
6. What are mirages?
7. How are rainbows formed?
8. What colours form yellow?
9. What colours form cyan?
10. What colours form magenta?

Light and Optics: Part 7: Answers

1. This states that the angle of reflection is equal to the angle of incidence.
2. This is a thin beam of light.
3. This means complete shadow. The light is completely blocked by the object.
4. This means part light, part shadow.
5. This is the minimum possible angle of incidence in which a total reflection occurs when light travels from a medium of higher refraction to a medium of lower refraction.
6. They are optical illusions or effects caused by weather or atmospheric conditions.
7. Rainbows are formed by the reflexion and dispersion of sunlight in water droplets.
8. Green and red combined.
9. Blue and green combined.
10. Red and blue combined.

Light and Optics: Part 8: Questions

1. What is phosphorescence?
2. What is fluorescence?
3. What is a light year?
4. Name the 3 primary colours.
5. What is dispersion?
6. What kind of image is formed by a diverging lens?
7. What is frequency?
8. What is focal length?
9. What does the speed of light depend on?
10. What is the speed of light in a vacuum?

Light and Optics: Part 8: Answers

1. It is a body that emits light a little longer after being irradiated by a light source.
2. It is a body that emits light after being irradiated by a light source but does not emit anymore light when the source of light is no more.
3. This is the distance travelled by light through a vacuum in a year.
4. Red, green and blue.
5. This is the separation of light by frequency into colours.
6. A virtual image.
7. This is the number of vibrations per second.
8. This is the distance between the focal point and the lens.
9. The speed of light depends on the surface it travels through.
10. 300,000 km per second.

Light and Optics: Part 9: Questions

1. What is an opaque object?
2. Give an example of an opaque object.
3. What is a translucent object?
4. Give an example of a translucent object.
5. What is a transparent object?
6. Give an example of a transparent object.
7. What is the shadow zone?
8. How does light travel in a homogeneous and transparent environment?
9. Name a natural ultraviolet wave source.
10. What protects us from the UV of the sun?

Light and Optics: Part 9: Answers

1. It is an object that does not allow light to pass through.
2. Concrete.
3. It is an object that does not allow light to pass through in a straight line.
4. Ground glass.
5. It is an object that allows light to pass through in a straight line.
6. A glass.
7. This is the part of an object that does not receive light.
8. It travels in a straight line.
9. The Sun.
10. Ozone.

Light and Optics: Part 10: Questions

1. What is a light source?
2. Name the two types of light sources.
3. Give examples of a primary light source.
4. What is incandescence?
5. Give an example of incandescence.
6. What is luminescence?
7. Give an example of luminescence.
8. What is a primary light source?
9. What is a secondary light source?
10. Give an example of a secondary light source.

Light and Optics: Part 10: Answers

1. It is a body that emits light.
2. A primary source and a secondary source.
3. The sun, the stars, a bulb, etc.
4. This is when a body emits light when it is heated to a certain (high) temperature.
5. Lightning.
6. This is when a body emits light at an ambient temperature.
7. A fluorescent tube.
8. It is a source of light that creates and emits light.
9. It is a source of light that reflects light created by a primary source.
10. The moon.

Light and Optics: Part 11: Questions

1. What is the point of incidence?
2. What is the incident ray?
3. What is the angle of incident?
4. What is the angle of reflection?
5. What is the reflected ray?
6. What is the normal of the object?
7. What is the plane of incident?
8. What is the refractive index?
9. What is the relation between the refractive index and the speed of light?
10. What is total reflexion?

Light and Optics: Part 11: Answers

1. This is the point struck by the light beam.
2. It is the light ray that strikes on the object.
3. It is the angle formed by the incident ray.
4. This is the angle formed by the reflected ray.
5. This is the light ray that is reflected from the object.
6. This is the perpendicular to the plane of the object.
7. The plane of incident consists of the incident ray, the normal ray and the reflected ray.
8. This is the speed of light in a vacuum (medium 1) on the speed of light in the second surface (medium 2).
9. The speed of light is low when the refractive index of a medium is large.
10. This is when all the light is reflected, so there is no refraction.

Light and Optics: Part 12: Questions

1. What is the link between the angle of incidence and the angle of reflection?
2. What is a real image?
3. What is a virtual image?
4. Give an example of a virtual image.
5. Give an example of a real image.
6. What are the characteristics of a converging lens?
7. What are the characteristics of a diverging lens?
8. Is the image in a convex mirror upright?
9. Give another name for a convex lens.
10. Give another name for a concave lens.

Light and Optics: Part 12: Answers

1. The angle of incidence is equal to the angle of reflection.
2. A real image is an image where light rays converge.
3. A virtual image is an image where light rays do not converge.
4. A mirror image.
5. A camera image.
6. It has a real image and its image can be received on a screen.
7. It has a virtual image and its image cannot be received on a screen.
8. No.
9. A converging lens.
10. A diverging lens.

Light and Optics: Part 13: Questions

1. What is presbyopia?
2. What is the cause of presbyopia?
3. What kind of lenses are used to correct presbyopia?
4. What is astigmatism?
5. What is myopia?
6. What kind of correction lenses are used for myopia?
7. What is hyperopia?
8. What kind of correction lenses are used for hyperopia?

Light and Optics: Part 13: Answers

1. This is when the crystalline lenses become rigid.
2. This is due to aging.
3. Convex lenses.
4. This is a blurred or distorted vision.
5. This is when someone sees near objects clearly but not objects that are far.
6. Concave lenses.
7. This is when someone has difficulties seeing near objects.
8. Convex lenses.

Pressure, temperature, heat: Part 14: Questions

1. What is temperature?
2. What is atmospheric pressure?
3. What instrument is used to measure the atmospheric pressure?
4. What can influence pressure?
5. What is the formula of pressure?
6. What is Archimedes force?
7. What influences Archimedes' force?
8. What is the relationship between the density of a liquid and Archimedes' force?
9. What is the relationship between the volume of a submerged liquid and Archimedes' force?
10. The pressure of a liquid depends mainly on?
11. Is there Archimedes' force in gases?

Pressure, temperature, heat: Part 14: Answers

1. This is what indicates the hotness or coldness of an object.
2. It is the pressure of the air that surrounds us.
3. A barometer.
4. The surface of contact, the larger the surface, the lesser the pressure.
5. It is force applied over the surface.
6. It is the force that opposes the force of gravity in a liquid.
7. The volume of an object and the density of the liquid.
8. As density decreases, the force of Archimedes decreases.
9. As the volume of the immersed liquid increases, the force of Archimedes increases.
10. The height of the liquid.
11. Yes.

Pressure, temperature, heat: Part 15: Questions

1. What is specific heat?
2. What is thermal expansion?
3. What is linear expansion?
4. What is thermal contraction?
5. What is latent heat?
6. How is heat transferred?
7. What is conduction?
8. What is convection?
9. What is radiation?

Pressure, temperature, heat: Part 15: Answers

1. This is the characteristic of a body to store heat.
2. It is the increase of the dimensions of a body following an increase of its temperature.
3. It is the increase of the length of a body following an increase of its temperature.
4. It is the decrease of the dimensions of a body following a decrease of its temperature.
5. This is the heat required to change the phase of a substance.
6. By conduction, convection and radiation.
7. This is the transfer of heat by contact.
8. This is the transfer of heat when it moves away from its source of heat.
9. This is the transfer of heat by electromagnetic waves.

Mechanics: Part 16: Questions

1. What is mass?
2. A ton equals how many kilograms?
3. What is density?
4. What is force?
5. How is force represented?
6. How is the force of gravity measured?
7. The force of gravity is expressed in which unit?
8. What are the characteristics of weight?
9. How is mass measured?
10. Is friction a force?

Mechanics: Part 16: Answers

1. It is the quantity of matter. The mass of an object is always constant.
2. 1000 kg.
3. It is the mass of a body over its volume.
4. It is anything that can modify the state of a body, either its form or its movement.
5. It is represented by a vector.
6. It is measured with a dynamometer.
7. In Newtons.
8. Its direction, its intensity and its centre of gravity.
9. It is measured with a scale.
10. Yes.

Mechanics: Part 17: Questions

1. What is net force?
2. What is centripetal force?
3. What is centrifugal force?
4. What is resistance force?
5. What is momentum?
6. What is inertia?
7. What is an elastic collision?
8. What is an inelastic collision?
9. Name 2 types of friction.
10. What is kinetic friction?
11. What is static friction?
12. What is an interaction pair?
13. What is an impulse?

Mechanics: Part 17: Answers

1. This is the sum of all the forces acting on an object.
2. This is the force that keeps an object in a circular motion.
3. This is the force moving outward from the centre.
4. This is the force opposite to the movement of an object.
5. This is an object in motion.
6. This is when an object resists changes in motion.
7. This is a collision in which kinetic energy remains constant.
8. This is a collision in which there is loss of kinetic energy.
9. Kinetic friction and static friction.
10. It is the force that slows down the movement of an object.
11. It is the force that permits movement of an object.
12. This is a pair of forces equal in strength, but in opposite direction.
13. This is the force multiplied by the change in time.

Wave: Part 18: Questions

1. What is a wave?
2. What is frequency?
3. What is the unit of frequency?
4. What is a mechanical wave?
5. What is a transverse wave?
6. What is a wavelength?
7. What is a period?
8. What is a node?
9. What is an antinode?
10. What is a longitudinal wave?

Wave: Part 18: Answers

1. A wave is a disturbance that transfers energy.
2. This is the number of cycles in one second.
3. Hertz.
4. This is a wave that requires a medium for its transmission.
5. This is a wave in which its direction of displacement is perpendicular to its direction of propagation.
6. This is the distance between two points in a wave.
7. This is the time of a complete cycle.
8. This is a point in a standing wave where there is no displacement at each cycle.
9. This is a point in a standing wave where there is maximum displacement at each cycle.
10. This is a wave in which its direction of displacement is parallel to its direction of propagation.

Wave: Part 19: Questions

1. What is a constructive interference?
2. What is a destructive interference?
3. What is a crest?
4. What is a trough?
5. What is compression?
6. What is rarefaction?
7. What is diffraction?
8. What can cause diffraction?
9. Define subsonic.
10. Define supersonic.

Wave: Part 19: Answers

1. This is when waves of equal frequency and phase meet and add up to a larger amplitude.
2. This is when waves of equal frequency and opposite phase meet and cancel one another.
3. This is the point on a wave with the maximum value of upward displacement in a cycle.
4. This is the point on a wave with the minimum value of downward displacement in a cycle.
5. This is the region of a wave in which the pressure and density are higher.
6. This is a region in a longitudinal wave where the waves are spread out.
7. This is the change in the direction of a wave.
8. An obstacle or an opening, etc.
9. This is the speed below that of sound.
10. This is the speed above that of sound.

Wave: Part 20: Questions

1. What is periodic motion?
2. What is superposition?
3. What is a wave front?
4. What is amplitude?
5. What is a shockwave?
6. What is the speed of sound?
7. Can sound travel in a vacuum?
8. What is the sound frequency for infrasonic?
9. What is the sound frequency for ultrasonic?
10. What is the function of an oscilloscope?

Wave: Part 20: Answers

1. This is a motion that repeats.
2. This is when waves pass through each other.
3. This is a surface with a propagating wave that passes through all the points in the wave that have the same phase.
4. This is the measure of the size of a wave.
5. This is when a wave moves faster than the speed of sound.
6. 340m/s.
7. No.
8. Sound frequencies below 20Hz.
9. Sound frequencies above 20,000Hz.
10. It measures sound intensity.

Wave: Part 21: Questions

1. Name the two main types of waves.
2. Give examples of transverse waves.
3. Give examples of longitudinal waves.
4. What is doppler effect?
5. What is fundamental frequency?
6. What is intensity?
7. What is a pitch?
8. What is a standing wave?
9. Name different types of electromagnetic waves.
10. Where is X-ray used?

Wave: Part 21: Answers

1. Transverse and longitudinal waves.
2. Light waves, radio waves, string instrument example (the guitar).
3. Sound waves, pressure waves.
4. This is the change in wavelength and frequency between the source and an observer in motion.
5. This means the lowest frequency.
6. This is the power per unit area of a wave.
7. This is the level of a frequency.
8. This is a wave that seem motionless.
9. X ray, gamma radiation, radio waves, microwaves, radar waves, infra-red, ultraviolet and visible light.
10. In the field of medical radiography.

Wave: Part 22: Questions

1. Where is gamma radiation used?
2. Where are infra-red waves used?
3. Where are ultraviolet waves used?
4. What are infrasonic sounds?
5. What are ultrasonic sounds?
6. What is light intensity?

Wave: Part 22: Answers

1. In the field of radiotherapy.
2. In heaters.
3. In sunbathing salons. The lamps that tan the skin.
4. These are sounds that are too low to hear.
5. These are sounds that are too high to hear.
6. This is the brightness of light.

Radiation Protection: Part 23: Questions

1. Name some substances that emit radiation.
2. What is a nuclide?
3. What does a nuclide consist of?
4. What is a radionuclide?
5. What is the condition for a spontaneous transformation to take place for a radionuclide?
6. Give another name for a radionuclide.
7. What are isotones?
8. What are isobars?
9. Is the nucleus a solid object with a bounded surface?
10. What can be used to measure atomic mass?

Radiation Protection: Part 23: Answers

1. Uranium, radium, etc.
2. It is an atomic nucleus in an individual entity.
3. It consists of neutron and proton.
4. It is an unstable nuclide.
5. The final energy must be lower than the initial energy.
6. A radioactive nuclide.
7. Isotones are nuclides that have the same number of neutrons but different number of protons.
8. They are nuclides with the same mass number.
9. No.
10. Mass spectrometers.

Radiation Protection: Part 24: Questions

1. What is the unit of atomic mass?
2. What is the opposite reaction of nuclear fission?
3. What is the force that regulates the movements of atomic electrons?
4. The radioactive decay of radionuclide is what kind of process?
5. What is radioactive half-life?
6. What is the activity of a radioactive source?
7. What are the 3 categories of the disintegration modalities?
8. How are artificial radionuclides produced?
9. Where are unsealed radioactive sources often used?
10. Where are sealed radioactive sources often used?

Radiation Protection: Part 24: Answers

1. Uma.
2. Nuclear fusion.
3. Electromagnetic force.
4. It is a random process.
5. This is the time it takes to reduce a radioactive activity by a factor of 2.
6. This is the number of disintegration per unit of time.
7. When a nuclide is transformed to another nuclide. When a nuclide is transformed into several other nuclides. When the same nuclide is transformed into another energy state.
8. By activation, filiation and fission.
9. In nuclear medicine.
10. Irradiation in a general sense.

Radiation Protection: Part 25: Questions

1. What are the 2 main radionuclides created in the atmosphere?
2. Which is more numerous in the atmosphere, C-14 or C-12?
3. What is the utility of C-12 in the atmosphere?
4. What is ionization radiation?
5. Why are charged particles said to be directly ionizing?
6. Name some charged particles.
7. Name some uncharged particles.

Radiation Protection: Part 25: Answers

1. Tritium and carbon 14.
2. C-12.
3. The dating of plants.
4. It is a radiation capable of removing an electron from an atom.
5. This is because they interact continuously with the matter.
6. Protons, electrons, and alpha particles.
7. X-rays, gamma rays and neutrons.

Conclusion

Thank you once again for downloading this book. I hope it has been of service to you.

Please, if you enjoyed this book, I would like you to leave a review. It'd be appreciated.

Thank you.

www.ingramcontent.com/pod-product-compliance
Lightning Source LLC
Chambersburg PA
CBHW030530220526
45463CB00007B/2768